U0392443

Work 246

不需要词汇的语言

Language Without Words

Gunter Pauli

[比] 冈特·鲍利 著

[哥伦] 凯瑟琳娜·巴赫 绘

朱 溪 译

上海远东出版社

丛书编委会

主　任：贾　峰

副主任：何家振　闫世东　郑立明

委　员：李原原　祝真旭　牛玲娟　梁雅丽　任泽林

　　　　王　岢　陈　卫　郑循如　吴建民　彭　勇

　　　　王梦雨　戴　虹　靳增江　孟　蝶　崔晓晓

特别感谢以下热心人士对童书工作的支持：

匡志强　方　芳　宋小华　解　东　厉　云　李　婧

刘　丹　熊彩虹　罗淑怡　旷　婉　杨　荣　刘学振

何圣霖　王必斗　潘林平　熊志强　廖清州　谭燕宁

王　征　白　纯　张林霞　寿颖慧　罗　佳　傅　俊

胡海朋　白永喆　韦小宏　李　杰　欧　亮

目录

Contents

ZURI Learning Initiative

一只角蝉正看着一只鱿鱼在改变皮肤的颜色，然后又改变了条纹和斑点的形状。

"你可以操控自己的皮肤呈现不同的图案和颜色，"角蝉评价道。

A treehopper is watching a squid changing the colour of its skin, and then changing the shape of its stripes and spots.
"You manipulate your skin showing different patterns and colours," the treehopper remarks.

一只角蝉正看着一只鱿鱼......

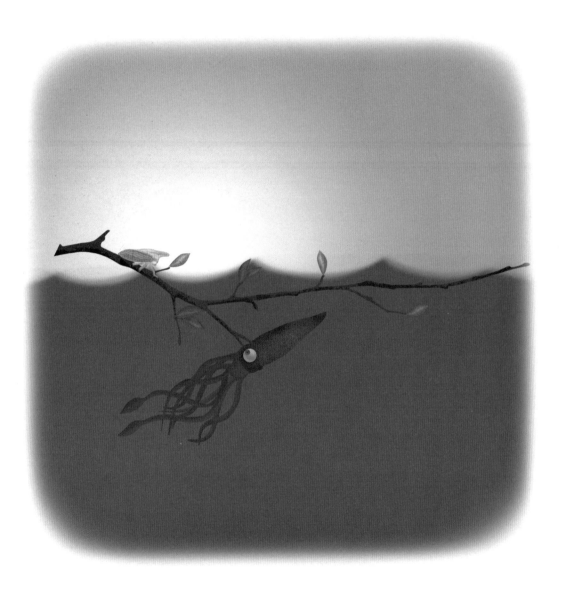

A treehopper is watching a squid ...

我们之间说话是通过振动......

Talk amoungst ourselves through vibrations ...

"我忙着说话呢，你没发现吗？你知道的吧，语言的源头并不是说出来的词句，"鱿鱼回答。

"我的确知道。我们角蝉之间说话也是通过振动。不同的震感可以让我直接向想要聊天的对象发送私人信息。"

"I am busy talking, don't you see? You should know that the origin of language was not the spoken word," the squid replies.

"I do indeed. We treehoppers talk amongst ourselves through vibrations. Different vibrations allow me to send a personal message directly to those I want to talk to."

"是的，在人类还没弄明白怎么说话之前，沟通就一直存在了。"

　　"一千年前，修道院里的修道士们使用手语互相交谈。你知道吗？那些修道士在喝啤酒以保持健康的同时一言不发地'交谈'了一千年。"

"*Y*es, communication has been around long before people figured out how to speak."

"*A* thousand years ago, people in monasteries used sign language to talk to each other. Do you know of those monks who have been 'talking' without words for a thousand years – while brewing beer to stay healthy?"

使用手语互相交谈……

Used sign language to talk to each other ..

... own sign language ...

"啤酒还能保持健康？只用手语交流？我看不出这两者之间有何关系……"

"人们了解到，在水中发酵谷物可以制成啤酒，而只要喝了它们，就不再会生病。因此，喝啤酒是为了庆祝身体健康，即使他们之间不用语言来交谈。"

"你看，连听障人士用自己的手语也有几个世纪了。"

"Beer to stay healthy? Talking in sign language only? I do not see what the one has to do with the other…"

"People learnt that by fermenting grains in water they can make beer, and if they drank this they would never get sick. So having a beer was a celebration of good health – even when they didn't talk to each other using words."

"Listen, the deaf have had their own sign language for centuries."

"而且我了解到，莫尔斯代码是由一位妻子失聪的男人发明的。一开始他在她的手上轻叩手指来发送消息。她能明白！后来他成名了。"

　　"人们试过很多方法来传递信息。贝尔先生的母亲也是听障人士。他想将声音转换成图形，以帮助聋人'看到'声音。"

"And I've learnt that Morse code was invented by a man whose wife was deaf. He started sending messages by tapping his fingers on her hand. She got it! And he became famous."

"People have tried many different ways to get their message across. Mr Bell's mother was deaf. He wanted to turn speech sounds into graphics to help the deaf 'see' the sounds."

在她的手上轻叩手指……

Tapping his fingers on her hand ...

将振动通过我们的双腿传递到……

sending vibrations through our legs ...

"看到声音？那不正是你在做的吗，鱿鱼先生？"角蝉问。

"是的，但他没有像我一样去呈现不同的线条和颜色，而是继续去发明电话了，却忘了与聋人说话的事。"

"我们角蝉通过快速弹动腹部来发展我们的通信系统，如此一来将振动通过我们的双腿传递到我们正坐着的植物的细枝上。"

"See sound? Isn't that what you are doing, Mr Squid?" Treehopper asks.

"Yes, but instead of making different lines and colours like I do, he went on to invent the telephone – and forgot all about talking to the deaf."

"We treehoppers developed our communication system by bouncing our bellies fast, and in this way sending vibrations through our legs, and into the twigs of the plants we are sitting on."

"你在显微镜下一定很可怕！放大后的你看起来可能就像雷龙一样发出刺耳的声音。真的一直都有效吗？"

"当我们改变振动时，便改变了我们说的话。你应该看看那些穿着白大褂的人类第一次偷听到我们的谈话时，他们的表情……"

"You must be scary seen under a microscope! Magnified you must look like a brontosaurus making squeaky noises. Does that really work all the time?"

"When we change the vibrations, we change what we say. You should see the faces of those people in white coats when they eavesdropped on our conversations for the first time…"

你一定很可怕……

You must be scary …

......在雨林和城市花园里。

... in rainforests and city gardens.

"我知道在雨林和城市花园里，有成千上万的角蝉整天忙于聊天度日。除非人们聚精会神，否则没人会注意到你们的。"

"要想'听到'我，你必须用所有感官去听才行，不只是用耳朵。"

"I know there are millions of you treehoppers, all busy chatting the day away, in rainforests and city gardens. No one notices you unless they concentrate carefully."

"To 'hear' me, you have to listen with all your senses, not just your ears."

"嗯，这让我想起了用鼻子'看'的海豚和用耳朵'看'的蝙蝠。现在多亏了你和我，生物交谈才增加了条纹和振动的方式。"

"这不就是自然之美吗？每天都有惊喜……"

"是的，只要每个人都准备好去聆听，即便什么声音也听不到！"

……这仅仅是开始！……

"Ah, that reminds me of dolphins that 'see' with their noses, and bats that 'see' with their ears. And now, thanks to you and me, we can add stripes and vibrations to the way creatures talk."

"Isn't that just the beauty of Nature? A surprise every day…"

"Yes, provided everyone is prepared to listen – without hearing a sound!"

... AND IT HAS ONLY JUST BEGUN!...

……这仅仅是开始！……

… AND IT HAS ONLY JUST BEGUN! …

Wild chimpanzees use at least 66 hand signals and movements to communicate with each other. Lifting food towards another chimp means "climb on me", while stroking its mouth means "give me the object".

野生黑猩猩至少使用66种手势和动作来相互交流。朝另一只黑猩猩举起食物意味着"爬到我身上来"，而抚摸它的嘴则意味着"把这个给我"。

Scientists have successfully taught apes more than one hundred words in sign language. At Columbia University in the USA, chimp named Nin was taught 125 signs.

科学家已经成功地教会猿猴上百种手语。在美国哥伦比亚大学，人们教会了一只名叫宁的黑猩猩使用125种手势。

右 左

Gestures involve the organisation of both hemispheres of the human brain. So, in order to understand the origins of language, we should study all cognitive properties and not just speech.

打手势涉及人的两个脑半球的组织配合。因此，为了理解语言的起源，我们应该研究所有的感觉特征，而不只是语言特征。

Babies learn to gesture at objects before they can speak, engaging the left side of the brain. This offers clues on how we structure spoken language and how it emerged in our ancestors.

婴儿在说话之前就学会对物体做手势，使用了左侧大脑。这为我们了解口头语言如何组构起来以及语言如何在先辈中出现提供了线索。

In sign language, words are conveyed by the hands. Intonation that is indicated by the rise and fall of one's voice is conveyed by facial expressions and different tilts of the headin sign language.

在手语中，单词是通过双手来传达的。手语中的面部表情和头部的不同倾斜度在口语中对应的是用声音的升降所表示的语调。

Separating and combining gestures and facial expressions provide the ability to communicate an infinite number of messages. This is a stepping-stone to the human language.

手势和面部表情的分离与组合使无数消息得以沟通。这是人类语言发展的垫脚石。

抹香鲸会通过声音来"说话"，构造音节，并根据它们漫游的海洋而有不同的表达方式。来自印度洋的抹香鲸所使用的语言与它们在大西洋的同类所使用的语言不同。

Sperm whales "speak" using sounds, construct syllables and have different expressions depending on the oceans they roam. Sperm whales from the Indian Ocean speak a different language to those of the Atlantic.

蝙蝠

蝙蝠会使用各种各样的鸣叫来相互交流。这些鸣叫包括 33 种不同的声音，或是蝙蝠单独使用的"音节"，或是以各种方式组合起来形成的"复合"音节。

Moustached bats use a wide variety of calls to communicate with each other. These calls include 33 different sounds, or "syllables," that the bats use alone, or combine in various ways, to form "composite" syllables.

Do you understand sign language? Would you like to learn?

你能读懂手语吗？你愿意学习手语吗？

Would you like to talk to a squid?

你愿意和鱿鱼说话吗？

How easy is it to read your dad's or mom's face?

读懂爸爸或妈妈的表情容易吗？

Would you like to have a surprise every day?

你想每天都有惊喜吗？

What is the difference between the language that humans use and the language all other living species use to communicate? Human language follows five guidelines: it is composed of small repeatable parts; it tells something about things that are not around; the whole has more meaning than the parts; there is no limit in length, and all words have special meanings. Your task is to demonstrate to your friends and family members whether the language that animals use follows the same guidelines or not, be it the language of whales, chimpanzees, treehoppers or squid.

人类使用的语言与其他生物交流的语言之间有什么区别? 人类语言遵循五个准则: 它由简短、可重复的部分组成; 它可以描述一些不存在的事物; 整体比部分更有意义; 没有长度限制, 而且所有单词都有特定含义。你的任务是向朋友和家人证明动物使用的语言是否遵循相同的准则, 可以举鲸、黑猩猩、角蝉或鱿鱼的语言为例。

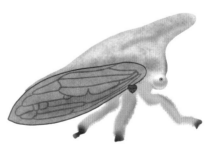

学科知识
Academic Knowledge

生物学	学习新语言有助于大脑发育；角蝉是树蝉和叶蝉的近亲；基因缺失是导致耳聋的原因之一；会变色的头足纲动物的皮肤。
化 学	第一台印刷机中使用的油墨是清漆（不是油墨）；鱿鱼的伪装可达到分子级别；色素体即能迅速延展成彩色圆点的具有弹性的色素囊；虹细胞反射所有碰到的光。
物 理	通过振动传递信息；海豚和蝙蝠可以回声定位，能用声音来"看"；闪烁的彩虹色，与颜色完美对称。
工程学	机器翻译是全自动计算机软件；光可以结合莫尔斯代码来传输数据；手语是具有自己语法的独特自然语言，不同于任何口语；逆向工程。
经济学	基于语言的归属感会对个人的社会经济地位和收入产生影响；一个国家的语言技能会被视为个人和社会可以投资获利的领域；双语顾客更喜欢使用自己母语提供的商品和服务；失去初始语言技能的移民子女可能每年都会遭受一定的收益罚金。
伦理学	语言不仅限于口语；无须倾听；学会的语言越多，就越容易了解他人。
历 史	语言学家得出的结论是，语言起源于公元前100000年左右。最早的木刻印刷的纸质书籍是创作于公元868年的中文书籍《金刚经》；欧洲的第一本书是由古腾堡于1455年印刷的；毕升在1000年左右发明了活字印刷术；第一套面向聋人教学的手动字母系统在1620年出版并在西班牙普及。
地 理	所有语言的三分之二来自亚洲和非洲；南非拥有最多的官方语言，多达11种；刚果的金沙萨是仅次于巴黎的世界第二大法语城市；美洲原住民利用简单的手势与其他部落进行交流，并促进与欧洲人的贸易。
数 学	数学语言结合了自然语言，技术术语和语法惯例，并辅以符号表示法；人工智能和图像识别。
生活方式	语言天才是指懂多种语言的人；发酵食品有助于释放营养元素以及保持健康；修道士会立下缄言誓约；喝含酒精的饮料时要举杯敬"健康"，因为酒精与细菌和病毒的控制有关。
社会学	全球约有7 000种语言；柬埔寨语字母表最长，有74个字母；罗托卡特的巴布亚语只有11个字母，是最短小的字母表；手势和肢体语言是最古老、最基本的交流形式；双语人士的大学入学率要高于双语能力较弱或没有双语能力的人。
心理学	语言心理学研究语言因素与心理因素之间的相互关系；不说话就能听懂的能力；沉默对我们思想的影响；联觉是指一种感觉或认知途径的刺激导致第二种感觉或认知途径的体验（看见声音或听见光）；每天都有惊喜的重要性。
系统论	健康、食物、饮料和沉默之间的关系。

情感智慧
Emotional Intelligence

角　蝉

角蝉善于观察。尽管身材矮小，但他并没有被体型更大的鱿鱼吓倒，并保持了自信的姿态。他明确表示自己能完美地以特有的方式，通过振动进行交流。他以近乎夸耀的方式分享智慧，在提供信息的同时通过提问来测试鱿鱼的智力。当鱿鱼不了解某些事情时，角蝉会给出一些令人惊讶的回答。角蝉通过提问题来吸引鱿鱼，这让他有机会解释自己及同类如何与他人交流。角蝉吹嘘他们的表现让科学家震惊。渺小的角蝉赞颂大自然的奇观，表明虽然他们微不足道，却有许多值得分享的特质，而这些是被世人所忽略的。

鱿　鱼

鱿鱼认为，每个人都知道她皮肤上变化的图案是其交流的方式，这是不言而喻的。她很自信地把人类语言置于历史背景之下，即许多其他物种在人类出现之前就具有良好的交流能力。虽然鱿鱼非常聪明，但是当受到角蝉的质疑时却无法将事实建立联系。面对角蝉能提供新信息的能力时，她想通过提出有关电话发明的详细信息来与之匹敌。关于海豚和蝙蝠的对话以一种轻松的方式继续进行，鱿鱼则添加了哲学性的评论：我们必须准备好去倾听。

艺术
The Arts

观察角蝉的外壳，它有哪些形状？有什么颜色？这么小的昆虫怎么会有这么大的创造力？让我们也发挥创意。找一些旧报纸并弄湿。用潮湿的报纸条来制作你最感兴趣的三只角蝉的甲壳形状。只有用到某些形式的加固材料如金属丝，才能做出角蝉的部分肢体。学习用金属丝和制型纸板手工模拟角蝉的形状和结构。等作品干燥之后再上色。

思维拓展
Systems: Making the Connections

关键是要聆听。我们每个人都有说话的能力，形成声音和音节组成的单词，再使用这些单词来构建表达含义的句子。但我们常常只专注于自己想说的话，而没有耐心聆听。我们都应该成为更好的倾听者。有很多方法可以表达自我以及传递希望分享的内容。人类创造的第一种非言语语言是手语，是一千年前在修道院中发展起来的。在艺术中，有着明确的规则来规定无需说话即可传达信息的方式，例如通过面具、面部表情和哑剧。原始绘画可以通过符号表达含义。为了与视障和听障人士进行交流，人们开发了盲文和莫尔斯电码等新技术。直到最近我们才意识到，除了口头语言，还有许多其他交流方式。越来越明显的是，我们过去看待语言是以人类自我为中心，这让我们难以理解自然界中各种事物之间不断交流的程度。现在我们了解了世界不仅拥有丰富的物种多样性，而且还拥有沟通的多样性。自然界中各种各样的交流方式和方法，超出了我们的想象，甚至超越了我们定义的语言。就我们人类而言，我们应该学会认真聆听。我们还应该关注生态系统中其他生命形式之间发生的多种多样的交流方式从而学习发现更多。不是每个人都说相同的语言，也不是每个人都以相同的方式讲话。这种意识将拓展我们的视野，让我们更加尊重与我们共享这个星球的所有物种。

动手能力
Capacity to Implement

无人能理解的语言有什么用？人们会说的语言越多，世界的体验就会越丰富。看看哪些语言可供你学习。它不必是常见的语言，可以只是口语。和亲朋好友聊聊，看看谁有兴趣与你一起学习新的语言。组织一门大家都可以参加的课程。关键是要在分享中学习。你有可能开启了一个新趋势。

故事灵感来自
This Fable Is Inspired by

卡罗尔·迈尔斯
Carol I. Miles

卡罗尔·迈尔斯在美国西雅图的华盛顿大学获得学士学位和博士学位。迈尔斯博士是美国纽约宾汉姆顿大学生物科学副教授，负责研究行为的神经基础。她一直领导有关角蝉振动通信的神经和机械基础研究。她的研究方法是首先了解昆虫的行为，探索昆虫行为在其一生中如何随着环境变化而调整。然后，她使用电生理方法探索行为的神经基础及其调控。她的研究包括角蝉进行振动交流的神经和机械基础，以及在寄生的烟草天蛾幼虫中观察到的进食行为和前馈活动变化的神经基础。

图书在版编目（CIP）数据

冈特生态童书.第七辑：全36册：汉英对照 /
（比）冈特·鲍利著；（哥伦）凯瑟琳娜·巴赫绘；
何家振等译.—上海：上海远东出版社，2020
ISBN 978-7-5476-1671-0

Ⅰ.①冈… Ⅱ.①冈… ②凯… ③何… Ⅲ.①生态
环境–环境保护–儿童读物—汉英 Ⅳ.①X171.1-49

中国版本图书馆CIP数据核字（2020）第236911号

策　　划　张　蓉
责任编辑　程云琦
封面设计　魏　来　李　廉

冈特生态童书

不需要词汇的语言

[比]冈特·鲍利　著
[哥伦]凯瑟琳娜·巴赫　绘

朱　溪　译

记得要和身边的小朋友分享环保知识哦！
八喜冰淇淋祝你成为环保小使者！